Plants and Flowers
of the Desert

Longman · York Press

All plants need water to grow.
Where there is sunshine and plenty of water
plants can grow well.

In the desert the hot sun shines day after day.
But there is almost no water.
Few plants can grow here.

3

But some plants can grow well
with very little water.
They can grow in the desert.

4

Not all of them are tough and dry.
Some desert plants have beautiful flowers.
How can they grow in the desert?

This tamarisk tree has very long roots.
It finds water deep under the ground.

6

It is not always dry in the desert.
From time to time it will rain.
This grass has long roots just under the sand.
They catch the rain before it dries.

Some plants can store water in their stems.
They can live without rain for many months.

This is the biggest plant that stores water
in its stem.
It grows in the American desert.
It is called a saguaro cactus.

This plant stores water under the ground
in a big round stem called a bulb.

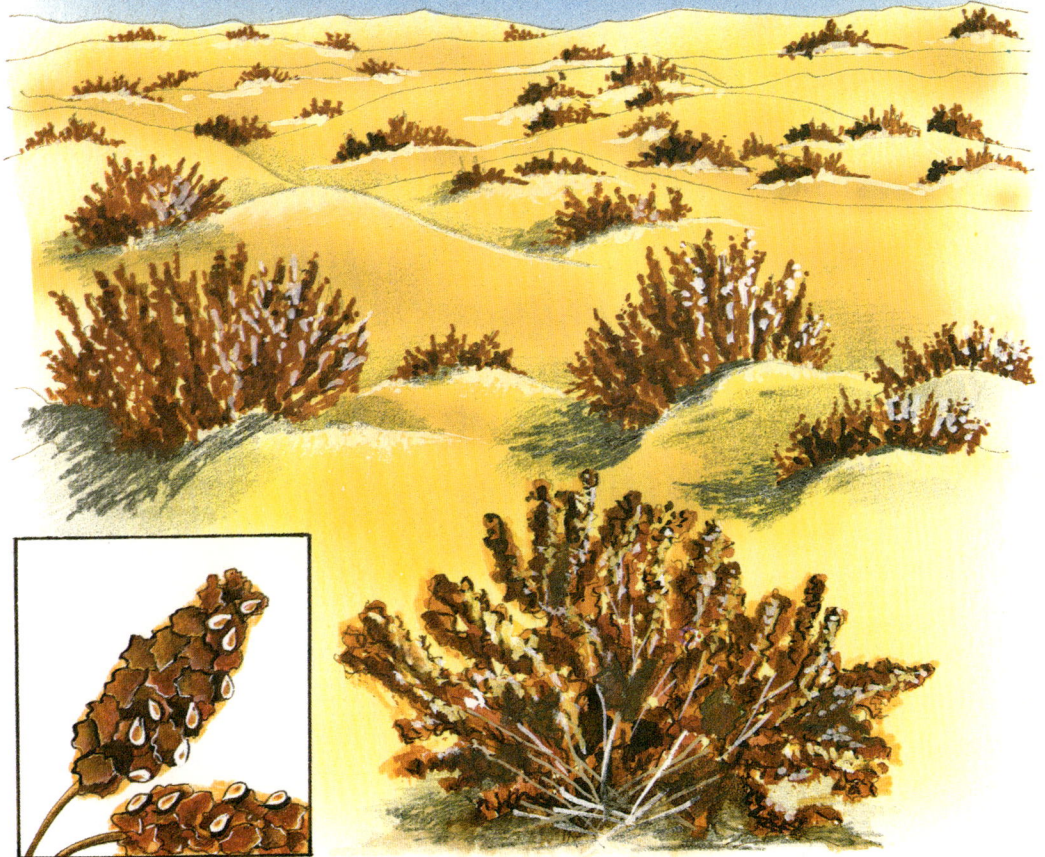

Sometimes in the early morning
you can see drops of water on plants.
This is called dew.
Some plants feed from the dew.

When it is very hot some plants
drop their leaves to save water.

12

These plants live in cracks between rocks.
The rocks give them shade from the sun.

In some places in the desert
it does not rain for many years.

But when it does rain
hundreds of plants grow very quickly.
How does this happen?

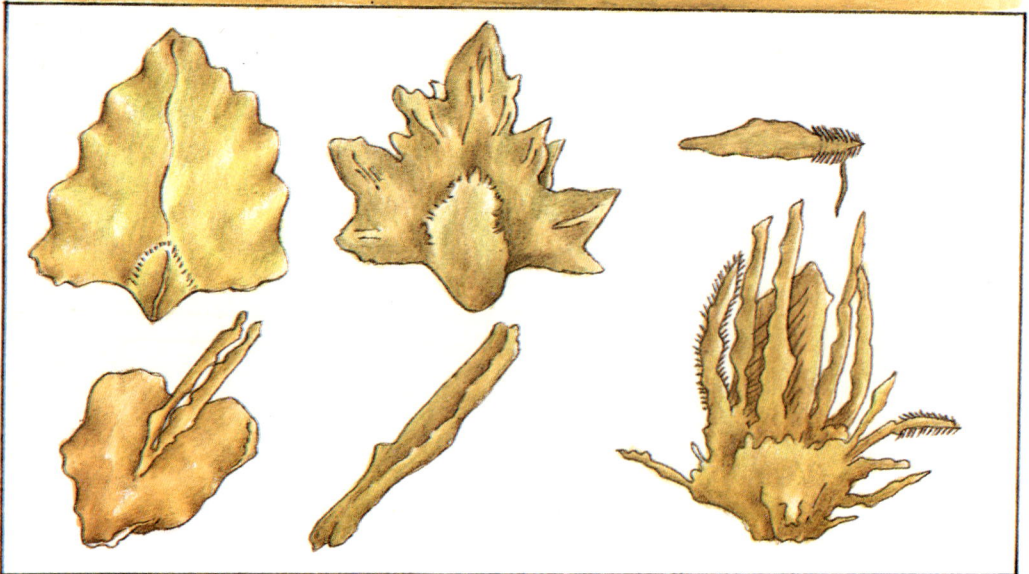

The seeds of some plants lie in the
ground for many years.
If there is no rain, they do not grow.

When it rains they must grow very fast,
before the rain dries up.

Two or three weeks after the rain
the plants have beautiful flowers.

These plants may live for only one month,
but they have already made more seeds.

Some seeds travel a long way
before they find a good place to grow.
These wild melons are blown along in the wind.
The seeds are inside the melon.

Other seeds remain in the same place
and wait until the next shower of rain.
This plant buries its seeds in the sand
so that they do not blow away.

There are so few plants in the desert
that many animals eat whatever they can find.
This oryx is eating a pretty desert flower.

This pretty cactus has sharp spikes
to keep the animals away.
You can eat the fruit from this cactus
but you must be careful how you pick it.

No animals like to eat this plant.
It will make them sick.
But it is very pretty.
It is called an oleander.

But animals can also help some plants to grow.
There are little spikes on these seeds.
The seeds stick to the fur of animals.
They may fall off where the animal finds water.

These goats are eating the bushes and grass.

When the goats have gone
there are very few plants left.

The roots of plants hold down the sand.
When there are no plants
the wind can blow the sand about.
The desert becomes very, very dry.
No plants can grow there then.

Grass with long roots can be planted
to stop the sand moving.
Then plants can grow again.

Life in the desert is hard.
People must take care to protect the plants.
Otherwise life in the desert
will become even harder.

Did you know that . . .

Some desert plants grow very slowly.
One cactus called the 'living rock' cactus
only grows about 1 centimetre every 10 years.

The roots of some shrubs in the American desert
can grow to over 560 kilometres in length.
The millions of small roots of one plant
can take in water over an area of 100 hectares.

The saguaro cactus is the largest in the world.
One saguaro cactus in New Mexico was discovered
to be over 16 metres tall
and to weigh about 10,000 kilogrammes.

The leaves of the acacia tree are almost half water.
These leaves provide both food and drink for gazelles.

The wild fig tree in South Africa has roots
which grow down 120 metres into the ground.

If plants grow too close together, some plants can
give out poison to kill their neighbours.
This allows them to use all the water themselves.

Index

LONGMAN GROUP LIMITED, Burnt Mill, Harlow, Essex
YORK PRESS, Immeuble Esseily, Place Riad Solh, Beirut
Illustrations by Fiona Almeleh

© Librairie du Liban 1985

First published 1985   ISBN 0 582 24507 9
Printed in Spain